The EQ and Compression Handbook

Learn the step by step process
to use EQ and Compression together

Nathan Nyquist

The EQ and Compression Formula:

Learn the step by step way to use EQ and Compression together

Copyright © 2017 Nathan Nyquist. All rights reserved.

Protected by copyright laws of the United States and international treaties.

No part of this publication in whole or in part may be copied, duplicated, reproduced, or transmitted in any form or by any means, electronic or mechanical, including photocopying, recording, or by any information storage and retrieval system, without the express written permission from the publisher.

Copyright and other intellectual property laws protect these materials and any unauthorized reproduction or retransmission will constitute an infringement of copyright law.

Federal law provides severe civil and criminal penalties for the unauthorized reproduction, distribution, or exhibition of copyrighted materials. Penalties for criminal and statutory copyright infringement are set forth at 18 U.S.C § 2319.

ISBN: 9781980601807

Table of Contents

Introduction 1

Beginning With the End in Mind 3

Part I: Equalization 7

The Front and Back Mixing Paradigm 8

What is EQ? 10

4-Zone Mix Theory 12

Broad Strokes First 14

The Rule of 300 16

EQ Controls Explained 18

EQ in Action 20

Slingshotting 22

The Real Secret to Mixing Nobody Told You 24

Frequency Sweeping 26

Part II: Compression 27

The 4 Truths About Compression 28

Compression Controls Explained 30

The Best Way to Learn Your Compressor 36

When and How to Use Compression 40

Using EQ and Compression Together 44

The Reality of Compression 46

Reaching The Next Level 50

Introduction

This book is designed to remove any confusion you might have surrounding EQ and Compression.

It's helpful to use this book as a reference guide, and frequently come back to it whenever you need guidance with EQ and Compression.

EQ and Compression can be a bit of a dry topic to learn and I believe that's part of the reason people have difficulty learning them. At the end of the day we're all artists and we just want to make good music, so it's important to master these tools as quickly as possible.

My goal with this book is to give you some simple paradigms and strategies that you can use to achieve a clear and well-defined mix.

2 | THE EQ AND COMPRESSION FORMULA

Beginning With the End in Mind

It's important to begin with the end in mind when mixing. Without a definite destination we are prone to wander aimlessly.

The purpose of a definite and clear goal is **NOT ALWAYS** to achieve it 100% exactly as you've set it. The point of a definite goal is that it gives you access to a powerful layer of subconscious resources and decision making which most of the time you're not aware of consciously.

It's like when you're at the cupboard and something begins to tip and fall over and your hand with the surprising reflexes of a ninja automatically reaches out and catches it.

We know that we wanted to save ourselves the mess; the effort of cleaning it all up.

So when we know what we want clearly enough, then reaching our goals becomes more of a reflex. It becomes a reflex because that's our unconscious taking control and making us do the things that would make it happen anyway.

So now that you're beginning to understand a little more about the benefit of clearer goals, let's talk about why you're reading this book.

4 | THE EQ AND COMPRESSION FORMULA

The fact that you're reading this means you probably want a well-defined mix or at the very least want to improve your mixing.

In order to achieve this, we need to know what a well-defined mix means to us.

The first thing to realize is great mixes are subjective. This means that to find a great mix, all we have to do is go find a few of our favorite songs. Then decide which of those songs you enjoy the 'Sound' of the most.

Whichever song you like the sound of most will become your **Reference Track.** This reference track will be your definite goal for a well-defined mix.

Your definite goal will be to match or exceed the sound quality of your reference track as much as you feel capable. If it seems you can't match or exceed the quality of your chosen reference track, that's ok because mixing is a gradual process of improvement.

All the most successful producers and mix engineers improved by referencing their favorite songs and paying close attention so they could eventually replicate that sound in their own music.

Most producers do this by keeping a copy of their reference track in their DAW session and listening back and forth between their track and the one they're matching.

6 | THE EQ AND COMPRESSION FORMULA

Part I: Equalization

The Front and Back Mixing Paradigm

Your mix can be thought of as having two spaces where you can place your instruments.

The two spaces are the **Front** and the **Back** of your Mix.

When mixing, you want to decide whether an instrument will be positioned up **Front,** which means it sounds like it's close to the listener or if it's positioned in the **Back,** which means it sounds like its far away.

The best way to really sensitize yourself to whether things in your mix are close or far away is to listen with your eyes closed. This is because when our eyes are open they tend to dominate our perceptions and actually decrease the sensitivity of our hearing.

Here are some simple distinctions I've made between the **Front** and **Back** of my mixes:

Instruments that are up **Front**:

1. Have more volume (louder compared to instruments in the back).
2. Have more high frequency content (compared to instruments in the back).

3. They tend to have less reverb because things that have lots of reverb get pushed further back than things which don't.

Instruments that are in the **Back**:

1. Have less volume (**quieter** compared to instruments up front).
2. Have less high frequency content (compared with instruments up Front)
3. Tend to have more reverb than the instruments up front.

You're mixes will instantly gain more depth and clarity when you make a practice of being definite about whether you want something in the **Front** or in the **Back**.

What is EQ?

EQ is a shaping tool. In the same way a sculptor use a chisel, in the same way women use makeup to change the shape and appearance of their faces, we use EQ to shape and chisel the presentation of our instruments.

Sometimes we're not only altering the shape of a sound, but instead correcting the tonal imbalances within a particular instrument or sound.

You might have an instrument with too much bass or too little treble. In this situation you would use EQ to correct these imbalances.

So EQ has two primary uses:

1. To sculpt and shape the presentation of our instruments.
2. To control the balance between instruments.

To achieve this result there are two main approaches to EQ:

1. Cutting EQ
2. Boosting EQ

Your primary mission with EQ will be to create a pleasurable sounding balance between instruments. In order to achieve that, we want to follow our **Front and Back Paradigm.**

By deciding which sounds are going to be in the **Front** and which ones will be in the **Back** of your mix you'll immediately become more exacting with your EQ decisions.

In the following sections I will show you the specific ways to use cutting and boosting EQ to achieve a proper **Front** to **Back** mix, but before we dive into that we need to understand '4-Zone Mix Theory.'

4-Zone Mix Theory

4-Zone Mix Theory refers to the four main regions of sound we will be shaping and sculpting within our song.

Understanding these 4 zones allows us to figure out which instruments are competing for any given frequency zone and then make EQ decisions to only those instruments as we improve the frequency balance within just that frequency zone.

The 4 zones are as follows:

Lows: 20hz - 200hz: Bass, Kick and Snare located here.

Mids: 200hz – 1khz: The meat and body of most instruments lies in this range. It's important to be very selective about which instruments you allow to be dominant within this range or else you'll get a muddy mix.

High Mids: 1khz – 5khz: This is where the forwardness and presence of your instruments reside. This is our ears most sensitive range (in particular our ear is most sensitive to boosts and cuts @ 3khz)

Highs: 5khz – 20khz: This is where the Sizzle, Aliveness, High-Definition quality of our instruments comes from. If the mix is too sibilant, this is the range too look at.

When we start cutting and boosting with EQ we will be focused on making what are called <u>Broad Strokes</u> in these <u>4 frequency zones</u> to achieve a well-defined **Front to Back Mix**.

Broad Strokes First

It's simpler and more intuitive to start with broad approach because mixing is a gradual process of sculpting and refinement. It's more effective to start broad and then funnel down to something specific.

If at the broad strokes stage you can't get 80% of the mix you're after, then you just haven't made the right broad strokes and small precision type strokes aren't going to fix your problem.

When we're mixing with broad strokes we're going to be relying on **WIDE** EQ Cuts and EQ Boosts where our Q= 1 or less.

By using wide EQ boosts/cuts we can then decide where we want to boost or cut by sliding through each of the 4 frequency zones and deciding which one yields the greatest improvement to our sound.

You will see an immediate improvement in the clarity and definition of your mixes by beginning with a broad strokes approach.

The Rule of 300

One of the most frustrating issues many producers encounter is a muddy or undefined low end in their mixes. Producers run into this problem because the instruments occupying the 20hz to 400hz frequency range of their mix are having a conflict catastrophe.

The reality is 20hz – 400hz is the most difficult frequency range for us to master because our hearing isn't as adapted to noticing details in it.

In order to avoid this temporary, but inherent weakness in our low end hearing there is one simple rule we can follow—The Rule of 300.

The rule of 300 states that if it isn't bass, kick or snare, then you must high pass those instruments at 300hz or higher.

The higher you can get away with the better for your mix.

This rule eliminates the main causes of a muddy mix because as you now know we run into this problem when our 20hz to 400hz range is swamped by too many competing instruments.

When professional mixers talk about balance they mean it as if there's a balance scale in front of you and so imagine on the one end there is something heavy, and the other has something light. And like the balance scale you must decide and **be very selective about which instrument(s) are frequency 'heavy' and which ones are frequency 'light.'**

Balance just means that it sounds pleasing to you, but when mixers talk about balance they mean that you're decisive about which instruments get to be frequency dense and which one's get to be frequency light in a given frequency zone.

By following this rule your mixes will immediately start occupying the top 5% of mixes out there.

EQ Controls Explained

High Pass Filtering (abbreviated HP): is used to remove unwanted low frequencies. It's something you'll use on virtually every track in your productions; especially because of the rule of 300 we just discussed.

Low Pass Filtering (abbreviated LP): is the opposite of high pass filtering. Generally low passing is used to remove unwanted high frequencies aggressively.

Cutting/Boosting (aka Peak Filtering): is when you take a bell shaped filter and use it to either subtract or add frequencies to a selected range.

Shelving Filters: are the ultimate broad strokes tool. I look at shelving in terms of whether I want to make something Brighter/Darker or Thicker/Thinner.

With a high-shelving filter set at 1khz or higher, I can easily make something Brighter or Darker sounding.

With a low-shelving filter set at 1khz or less, I can effortlessly make an instrument Thicker or Thinner sounding.

Unfortunately, the ease, effectiveness and power of these filters, cause them to be one of the most overlooked tools at a mixers disposal.

Q: is best understood as a way to adjust the sharpness of EQ curve. Different curves will produce different sounds. An interesting behavior is that sharper curves tend to be less transparent when boosting than their cutting variants.

EQ in Action

You probably already know on some level that mixing is mostly a subtractive art form. And so any good mixer will be using cuts about 80% of the time and boosts roughly 20% of the time.

The rule of thumb I hope you'll find useful is it's much better to cut too much from everything and then be very selective about where you restore fullness, than to not cut enough and get a mediocre and mucky mix.

Once you've got a lot of open space to work with, it's much easier to add frequencies back in, than it is to fit another person into a crowded bus so to speak.

There are 2 main techniques I rely on when I'm EQing my mix. As I've said before, I prefer to approach EQing in broad, even sloppy terms. It isn't necessary to be misled and think there is only one right way to EQ something, there are typically a few, and it's your job to find one quickly so that you can move on with it.

When I say quickly I mean that EQing is actually a messy and gradual process of refinement. That means it's really about little by little, sculpting and shaping our music into something we're gradually becoming more and more pleased with.

It's really helpful to understand that you're only making decisions as quickly as you're comfortable. And so make your decision comfortably fast because sticking in one place for too long risks over-analysis.

Over-analysis inhibits creative decision making and worst of all leads to increasingly worse decisions. That is why while mixing I like to use a technique I call **slingshotting**.

Slingshotting

Slingshotting works by using the 4 frequency zones we talked about earlier.

Here is the order of the technique:

Decide whether you want an instrument kept in the **Front** or pushed to the **Back** of your mix.

Then roughly set the desired volume you'd like your instrument to play at so it's in **front or back**.

With a Q slope of 1, select a 4db boost if you want to bring it more forward and a -6db cut if you want to push it further back.

Now, while everything is playing move your cut/boost between each of the 4 frequency zones. Notice in which zone you experience the most pleasing change.

Once you've found the zone you like most, then you can configure the db amount of your cut/boost so it's more refined.

The idea here is to provide you with only 4 different choices instead of so many choices you don't know which to select.

Keep in mind that cutting more is better than cutting too little. Boosting too little is better than boosting too much.

When you're searching for a zone to cut and you find the right range you'll notice that your mix literally opens up and feels less congested.

The goal is to aim for openness and almost too much space between instruments because **the real secret to a dynamite, 3-dimensional mix has to do with the space you leave unoccupied.**

The Real Secret to Mixing Nobody Told You

"The real art of mixing is about the "Frequency Real Estate" you don't use, not the space you occupy."

Really what I'm saying is less is more.

Why is less, in fact more? My belief is because of how limiting/mastering works. There *is in fact...* a good reason people don't notice how muddy their mixes are until they start limiting and mastering them.

It's because limiters aren't really loudness maximizers so much as they are intensity multipliers.

When we fail to aggressively cut frequencies we don't need, then instruments start competing for dominance. If there's too many instruments fighting for dominance in a given frequency range, then *when we go to limit our track we increase the intensity of those clashing frequencies*.

If our mixing were on point, then we would just be increasing the intensity of an open, clear, and definite mix which would sound absolutely wonderful.

So keep it clean, keep it nice and open.

Frequency Sweeping

Frequency Sweeping is similar to Slingshotting, only whereas Slingshotting is meant to give you a result quickly and definitively; frequency sweeping is more surgical and precise and as a result it's more focused and time consuming.

Because it's more surgical and precise, frequency sweeping involves a very narrow Q with around + or – 10db of gain.

By using a very narrow Q you're able to microscopically zoom in to the specific points of the sound where you might be hearing a ringing or pinging, something which typically sticks out and annoys you.

It's easier to find the offending frequency on EQ's which have a built-in spectrum analyzer since the offending frequency will tend to poke out like a sore thumb from the rest of the frequency spectrum.

Part II: Compression

The 4 Truths About Compression

> "If you don't know when to use it, then don't."
> -Elitist Internet Forum Asshole

That tired, old advice always made me wonder how anyone ever learned to use compression in the first place? I never liked hearing people say it, but really what they were trying to say is compression isn't what will make or break your mix.

We want to begin to think of compression as more like the icing on a 5 layer cake. The 5 layers are great by themselves but maybe we want some icing to top it off.

There are 4 perspectives I've used to understand what compression really is and how to use it:

1. Compression is just an **automatic volume fader**. This is literally what a compressor should mean to you. **It's not compression, it's an automatic volume fader.**
2. Anything a compressor can do, volume automation can do.
3. Compressors are designed to make loud sounds quieter.
4. A compressor is just a glorified **volume fader** (I really want to drive that one home).

Compression Controls Explained

At a basic level a compressor has 5 main controls which we can group into 2 groups that allow us to easily navigate and control the behavior of our automatic volume fader.

The 1st group is called our **Fader 'Triggering'**. It consists of our Threshold, Ratio, and Sidechain controls.

These controls tell our automatic volume fader *'when'* to move down, and *'how much'* volume to subtract when it does.

The 2nd group is called our **Fader 'Speed Properties'** It consists of our Attack, Hold and Release controls.

These controls tell our automatic volume fader **'how fast'** it should move when it is triggered by the incoming signal.

Fader 'Triggering'

The controls in this section allow our automatic volume fader to know 'when' and 'how much' to push down the volume of the incoming signal.

Threshold

Threshold tells the volume fader 'when' to start decreasing the volume of our incoming signal. Signals which rise above our threshold cause our automatic volume fader to move down, thus decreasing the volume of our sound.

Signals which remain beneath our threshold cause our automatic volume fader to do absolutely nothing.

Ratio

Ratio is 'how much' our automatic volume fader will move down in response to the incoming signal when threshold is exceeded. This means when the volume of the sound is louder than our threshold setting, then the ratio determines 'how far' down the compressor will move our automatic volume fader.

The higher our ratio, the more our automatic volume fader moves down in response to an incoming signal which goes above our threshold.

The lower our ratio, the less our automatic volume fader moves down in response to an incoming signal which goes above our threshold.

The difference between a high ratio and a low ratio is like the difference between gravity on earth vs. gravity on the moon.

When you jump on earth you get pulled down right away, but when you jump on the moon, you go higher and you get pulled down slower and more gently.

It's the same with our ratio control. The lower our ratio, the more gently our volume fader pushes down. The higher the ratio the more aggressively our volume fader pushes down.

Fader 'Speed Properties'

Attack

Attack is 'how long' it takes our automatic volume fader to move downward when the threshold is exceeded by our incoming signal.

So for example, with a 10ms attack when our incoming signal exceeds the threshold of our compressor, it means that our automatic volume fader takes 10ms to decrease the volume as determined by our threshold and ratio controls.

Once our attack is complete the volume fader will continue subtracting volume until the incoming signal gets quieter and falls beneath our threshold. Once this happens our automatic volume goes into the release phase.

Release is simply the time it takes for your automatic volume fader to move back up to 0 volume reduction.

Release

Release is 'how long' it will take our automatic volume fader to return back to its original position of 0 volume reduction.

The release phase happens only when our incoming signal has fallen beneath the threshold we've selected.

So with a 10ms release, it would take our lowered automatic volume fader 10ms to return back to its original level.

If as the compressor is completing its release timing the incoming signal goes above threshold, then the compressor will re-enter its attack phase until gain reduction is achieved again.

I know it's kind of redundant, but nobody ever explained what happens if your release isn't allowed to finish before the compressor is retriggered.

Hold

Hold is an extra parameter. Not all compressors have it because it isn't essential.

Hold actually happens before your release stage. **It's a way to delay the release phase** of our automatic volume fader.

So the order of these stages is:

Attack > Hold > Release

As you're aware it's during the release stage that the fader gradually returns back to its original position of zero volume reduction.

When we apply hold we are preventing ('holding') our fader from entering the release phase, which means we are maintaining our volume reduction at whatever point it's reached by the set duration of the hold parameter.

Whereas release is the time it takes our automatic volume fader to return back to its original position, hold is just hitting **pause** and keeping the fader stuck wherever it is until the incoming volume become quiet enough to fall beneath the threshold and thus trigger the release stage.

Hold just gives you another way to delay the return of your automatic volume fader to its original position.

Makeup Gain

Since our automatic volume fader functions by reducing the volume of the sounds, often the sounds we treat with compression can become quieter.

Makeup Gain allows us to restore our instrument to its original perceived volume level.

However there will be situations where you'll be using compression to only reduce the volume of a sound whenever it gets too loud.

This is because you're using the compressor as a volume sentinel who's only job is to viciously slap any overly loud sounds back into lower volume territory. This is one of the easiest and most natural ways to use your automatic volume fader.

Sidechain

The Sidechain of a compressor is used to control 'how the compressor responds to the incoming signal.'

A Sidechain's purpose is to allow the compressor to act as if it were behaving in response to an EQ'd version of the sound you're treating.

This is most often used to remove low frequencies from the incoming signal that's used to trigger the compressor behavior. It's useful since bass frequencies tend to cause compressors to respond more sluggishly and inconsistently.

Using the Sidechain to remove low frequencies makes our compressor behave more precisely and predictably.

The Best Way to Learn Your Compressor

The fastest way to learn any processor is to jump between extremes and then compare the results. This is how we will be learning compression.

Here's a step by step guide to learning what your compressor (automatic volume fader) does in action.

1. Pull a drum loop into your DAW. Ideally you want a drum loop that has room tone/ambience/reverb already on it.
2. Add a compressor to the drum loop track.
3. Play the loop so it's looping automatically.
4. As the loop is playing set the ratio to 10:1 and then move the threshold all the way down (this will quiet your loop considerably).
5. Now set the attack and release to 1ms or less (as fast as possible).
6. Gradually pull the threshold back up until you have 10 - 15db of gain reduction (volume reduction).
7. Play with your attack by opening it completely and closing it back down. Notice how the sound of the drum loop changes as you do this.
    a. You'll notice as the attack gets longer that the initial transient of each drum hit is allowed to poke through more and more
8. Set the attack back to its original position at 1ms or less.

9. Now play with your release in the same way by opening it and closing it. Notice how the drum loop changes as you do this.
    a. What you'll notice is that the slower the release, the more the tails of the drum hits and the room tone start to disappear.
10. Repeat steps 7 and 9 only now as you play with the attack/release also play with the ratio setting.
    a. You'll notice how higher ratios influences the aggressiveness and obviousness of your attack and release parameters.

This exercise will allow you to hear in a dramatic and focused way what the behavior of your automatic volume fader sounds like. In particular it sensitizes you to the rate and speed that your volume fader moves up and down.

What you'll notice is when the attack was slower, it allowed the drum attack or what we call transients to come through.

When you set the release so its slower and slower, you'll notice how the tails of the drum hits and the room tone start to diminish. That's because our automatic volume fader isn't allowed to return back to 0 volume reduction before the next drum hit triggers the automatic volume fader to quiet the sound again.

38 | THE EQ AND COMPRESSION FORMULA

This is the fastest and easiest way to learn the behavior and character of any compressor.

When and How to Use Compression

There are 2 primary ways I use compression on an individual, track based level:

1. To control and shape transients (attacks)
    a. Percussion shaping
    b. Percussive/Plucky Instrument Sculpting
2. To control the volume of a sound over time
    a. The vocalist who's volume is all over the place
    b. The guitar performance which is naturally very dynamic

Using Compression to Control Transients

1. Set your ratio 2:1.
2. Set both a fast attack and release of 1ms or less.
3. Pull your threshold down until your transient begins to disappear.
4. Now slow down your attack so that as much of the transient as you desire is allowed to come through.
5. Set your release timing between 10-40ms so the tail of your transient is relatively unaffected.
6. Adjust your Makeup Gain to return the sound to its original volume.

7. You can do a finishing pass by going back between your ratio, attack and release settings and tweaking them in conjunction with one another.

You can use this same process to **amplify and further shape the presentation of a transient**. The only difference would be to set a slower attack so the transient or 'head' is allowed to slip through unaffected before your automatic volume fader moves down to decrease the volume of the body sound.

Using Compression to Control the Volume of a Sound Over Time

I use vocals as an example because it's very easy to notice the effect of compression on vocals. This is because as human beings our hearing is naturally sensitive to the various audio cues present in the human voice.

1. Set your ratio 2:1.
2. Set your attack and release to 1ms or less.
3. Pull your threshold down until you have 10 db of gain reduction.
    a. With this amount of gain reduction a vocal will sound a little unnatural, wait until step 6 to fix that.

4. Now open your attack until the vocals consonants and plosives sound as pleasing and intelligible as you'd like.
    a. Very fast attacks settings will squash consonant/plosive sounds in a vocal. However, in a musical setting you can get away with faster attacks since music is creative you really can do whatever you feel like.
5. For natural sounding vocals set the release between 15 - 40ms.
6. Pull the threshold up until you have about 6db of gain reduction.
7. Adjust your makeup gain to return the sound to its original perceived volume level.
8. You can do a finishing pass by going back to your ratio, attack and release settings and further tweaking them in conjunction with one another.

Using EQ and Compression Together

There is an overall consensus that the majority of the time EQ should be placed before compression.

So, EQ > Compression

However there are no set rules and if you do find yourself EQing after you compress then the typical use in that scenario would be to cut with an EQ before your compressor and then boost with an EQ after your compressor

EQ (Cut) > Compressor > EQ (Boost)

When you become comfortable with compressors and EQs the above chain actually makes it easier to shape your sound even more dramatically.

But don't be misled, if all you mastered was the EQ > Compression combo, you would still be in the top 1% of mixers.

The Reality of Compression

The reality today is when I'm mixing a track I use compression on about 5-10% of tracks.

I tend to use compression in the same way a person might Febreeze™ their house after cleaning it. Febreeze™ is *really* just the icing on the cake, but only once you've cleaned everything is Febreeze™ actually worth it.

Otherwise you would have a house that looks like sh%t but smells like heaven.

Anyway, this is just how I look at compression: Only after I've already cleaned everything up with EQ, do I then use compression to put down the finishing touches.

If you're just learning compression and you don't know when to use it, then don't because I have a solution for that absurd catch 22.

Since we all need to learn compression at some point, I would suggest that you learn and familiarize yourself by using this simple approach.

All you're going to do is pick two instruments in your song, one in the **Front** of your mix and the other in the **Back**.

The goal here is to use compression to more effectively keep the **Back** instrument comfortably tucked in the **Back**, and to make the **Front** instrument stay in the **Front** more consistently.

I actually call this Front/Back style of compression, "Pocket Compression," because in the same way we use EQ's to carve out frequency pockets for instruments in our mix, we can use Pocket Compression to create a musically pleasing 'dynamic boundary' **AROUND** the frequency pockets we've carved with our EQ.

The reason creating this 'dynamic boundary' is amazing is because it allows us to achieve even greater levels of **blend/contrast.**

So even though we are always carving out pockets with EQ, those instruments can still sometimes get excited and jump up in volume. And as they jump in volume, they jump out of their pocket so to speak.

So you need something to make instruments stay in the comfortable pockets you carved out for them. This is where Pocket Compression comes in.

But here's the truth about pocket compression, because it's really just my way of using Bus Compression.

The reason I call it pocket compression is because it actually tells you the result you're getting. With pocket compression you're forcing instruments to energetically/dynamically stay in the same pocket.

In this particular form of compression you'll learn that you have frequency pockets which are managed by EQ's, but you also have volume-leveled pockets which are instead managed by Compression. It's a lot like our automatic volume fader from earlier, only now it's being applied to groupings of instruments instead of just one.

With Pocket Compression (aka Bus Compression), **you're forcing groups of instruments to sit in the same dynamic pocket** which has very distinct, time-regulated volume properties.

The particular 'time-regulated volume properties' of this pocket are controlled by your compressor. It's the particular configuration of the compressor for instruments in a given pocket which creates the characteristics of the Pocket you're Compressing.

Having just 2 of these distinct pockets in your mix means your mixes will have more depth, and therefore more **blend** and **contrast** than someone who doesn't use this technique. This allows us to create a powerful form of dynamic contrast in our mixes that is truly professional sounding.

Pocket Compression really is the most powerful and intuitive approach to compressor use that I know of and in my book

'The Bus Compression Masterclass' I expand completely on my philosophy and approach to this form of Bus Compression.

If you're curious and want to learn more about Pocket Compression then you can look up **'The Bus Compression Masterclass,'** on amazon.com.

Reaching The Next Level

I'm constantly experimenting to find that 'next level' and it's amazing because every time I do, I discover a simpler approach to getting an even better result.

The next level always gives me an easier, more guaranteed way to achieve a predictable result. Pocket Compression is one of those 'Next Level' paradigms.

And so as I continue evolving and getting better, all these older inefficient approaches get replaced by simpler, more intuitive tricks.

Reinventing the wheel isn't always necessary and since we're all artists, and we all want to get better at our art, then it's important we do everything to continue evolving.

The way I accelerate this process of evolution is by keeping a pen and notepad next to my studio computer and whenever I do something that I find glorious, I simply write down the steps I took to achieve it on paper.

This has the effect of fast tracking the learning process as well as making your automatic, reflexive access to skills and resources more robust and consistent.

I'm basically just keeping an archive of all my awesome engineering and production wizardry, because it would be cruel not to reuse the same wizardry-level successes until they become persistent reflexive successes in the future.

It would be so much nicer to reflexively catch ideas like we've caught falling objects from the cupboard than to go the other more difficult and unreliable way.

Keeping a notebook of my own techniques has provided an invaluable boost to the rate, speed and consistency of my own evolution and I hope you'll agree it's something that can help you too.

At the end of the day most of what's powerful is simple, ridiculously simple; especially compared with the complexity that was necessary build it.

But in a field like audio engineering that tends to attract the technically attracted, there's a trend towards drowning things in an ocean of complexity.

In the end art is meant to be the result of 'artistically' simplified complexity. And if we don't stretch to simplify, reduce, and intensify what remains then we will miss the benefits of challenging ourselves to always reach that next level.

At the highest levels everything is subtractive and simplified. Only what's absolute or essential remains, everything else is just clutter.

So if you're interested in learning more about audio engineering, music production and my philosophy of music you can go visit **www.mybeatlab.com** where you'll find free tutorials and content that will help you reach the next level.

Good luck and always keep stretching towards that next level.

Additional Resources

Get my New Book "The 3-Space Reverb Framework" at:
https://www.amazon.com/dp/B07BBX3FMN

Check out my advanced mixing book, "The Bus Compression Masterclass" at: http://amzn.to/2BXlwAP

Check out my EDM Bass Layering Masterclass at:
https://www.udemy.com/ultimate-bass-secrets/

Get free resources and tutorials at my website:
http://www.mybeatlab.com

Subscribe to my Youtube channel where I regularly release tutorial videos:
https://www.youtube.com/channel/UCX3RckO344GKk6F6ryj0bCQ